THE SCIENCE OF AI

AI Innovators

John Allen

ReferencePoint Press

San Diego, CA

About the Author
John Allen is a writer who lives in Oklahoma City.

Printed in the United States

For more information, contact:
ReferencePoint Press, Inc.
PO Box 27779
San Diego, CA 92198
www.ReferencePointPress.com

LIBRARY OF CONGRESS CATALOGING-IN-PUBLICATION DATA

Name: Allen, John- author
Title: AI Innovators/by John Allen- 1957
Description: San Diego, CA : ReferencePoint Press, Inc, 2027. | Series: The Science of AI | Includes bibliographical references and index.
Identifiers: LCCN 2026012934 (print) | ISBN 9781678213107 library binding | ISBN 9781678213114 ebook
Subjects: LCSH: Juvenile literature

For compete cataloging-in-publication data please go to www.loc.gov.

CONTENTS

Seeking AGI with DeepSeek

Few technology experts noticed when an obscure Chinese computer engineer used profits from his hedge-fund company to launch a venture called DeepSeek. The announced purpose of the new firm was to study artificial intelligence (AI). Two years later, on January 20, 2025, DeepSeek shocked the tech world by introducing R1, an open-source AI model with capabilities that rivaled OpenAI's ChatGPT for a fraction of its development cost. Panicked investors dumped shares of American tech stocks, setting off a temporary market drop of more than $1 trillion. All at once, America's supposedly safe lead in AI technology seemed in question.

But Liang Wenfeng, the chief executive officer (CEO) of DeepSeek and architect of all this turmoil, took the hoopla and controversy in stride. He was used to surprising people, ever since his student days in the Chinese province of Guangdong. At age seventeen he entered one of China's most elite universities and went on to get a master's degree in information and communication engineering. Today, after a rapid rise in China's tech industry, the youthful-looking Liang has a laser focus on developing AI at DeepSeek. His goal is to achieve artificial general intelligence (AGI), or autonomous systems that can outdo humans in performing valuable tasks. DeepSeek hires graduates from China's top universities, the kind of employees who love the challenges posed by advanced AI. "What attracts the best talent is obviously the solving of the world's hardest problems," says Liang. "Our goal is still to go for AGI."[1]

Building on the Work of Early Innovators

Liang is one of a group of innovators worldwide who are shaping the future of AI. They have helped bring about an explosion of AI breakthroughs, building on the work of scientists and engineers who first glimpsed the possibilities for intelligent machines. Among these early innovators was Frank Rosenblatt, a psychologist and computer pioneer at the Cornell Aeronautical Laboratory in Buffalo, New York. In 1957 Rosenblatt developed the first artificial neural network, which he called the Perceptron. It functioned as a single binary neuron—like a nerve cell in the brain—using learning algorithms to solve classification problems. The Perceptron's brain-like processes enabled it to learn by trial and error.

Over the next forty years, refinements of Rosenblatt's neural network model made it more effective and set the stage for more complex forms of machine learning. In the early 2000s the rise of

Liang Wenfeng is the founder of DeepSeek, a startup AI company that introduced an open-source AI platform at a fraction of the cost of its rivals.

the internet enabled computer systems to learn from huge datasets online instead of relying on human programming. Innovation in AI reached stunning new heights in the 2010s. So-called deep learning drew on Rosenblatt's ideas to create multilayered neural networks powered by computer chips called graphics processing units (GPUs). Then in late 2022, Sam Altman's OpenAI released ChatGPT, which uses large language models (LLMs) to generate text, images, and code on demand.

The emergence of generative AI owes a large debt to Rosenblatt's original brainstorm on neural networks. "Rosenblatt ventured into the unknown, breathing life into the first neural networks within the cold metal hearts of the 1950s computers in Cornell's Aeronautical Laboratory," says J. Christopher Westland, professor of information and decision sciences at the University of Illinois, Chicago. "[He] laid the groundwork for the neural network AI that would one day challenge the boundaries of human creativity."[2]

Turning Dreams into Everyday Reality

Like Rosenblatt, today's AI innovators are visionaries turning their dreams into everyday reality. Some, like Sam Altman and Jensen Huang, have become public figures who frequently meet with world leaders and whose views about AI make headlines. Altman's OpenAI, with its rollout of ChatGPT and later models, has helped make generative AI a practical tool for the masses. Huang, whose company Nvidia powers AI worldwide with its state-of-the-art GPUs, is helping bring the holy grail of AGI closer to fruition. Other innovators, like Demis Hassabis of Google DeepMind and Daphne Koller of the biomedicine company Insitro, are better known to industry insiders. Yet their work in using AI tools to analyze the root causes of disease and design new drug treatments promises to save lives. A major innovator such as Fei-Fei Li is also hardly a household name, even though her work with ImageNet has played a major role in training AI models to recognize images with remarkable accuracy. This ability is crucial for everything from face recognition in security software to self-driving cars. Li and other innovators are not only changing the AI industry, they are changing the world.

> **"We (the whole industry, not just OpenAI) are building a brain for the world. It will be extremely personalized and easy for everyone to use."[3]**
>
> **—Sam Altman, chief executive officer of OpenAI**

Altman in particular has stepped forward as a spokesperson for AI's potential. He sees each breakthrough as an exciting new baseline for future progress. "We (the whole industry, not just OpenAI) are building a brain for the world," he says. "It will be extremely personalized and easy for everyone to use. . . . For a long time, technical people in the startup industry have made fun of 'the idea guys,' people who had an idea and were looking for a team to build it. It now looks to me like they are about to have their day in the sun."[3]

CHAPTER ONE

Fei-Fei Li: Teaching AI to See

In 2007 a team of computer scientists at Princeton University were working on a way to train machines to see. The team, led by Fei-Fei Li (pronounced Fay Fay Lee), were compiling millions of images from the internet. Each image was assigned to one of thousands of categories. The dataset they were assembling became the largest collection of digital images then in existence. Yet the team's efforts were mostly met with skepticism from colleagues and other AI experts. They could not see the value of such a vast dataset. "I think you've taken this idea way too far," a mentor of Li's told her. "The trick is to grow with your field. Not to leap so far ahead of it."[4]

The logistics of the two-year project put a serious strain on Li's research budget—and the patience of her student helpers. Nonetheless, she refused to be discouraged. Her dataset project, called ImageNet, ended up revolutionizing AI by teaching it to see a more complex and diverse world. In the years since, Li has continued to seek practical applications for AI. She has also used her experience as a successful immigrant in the United States to support diversity and opportunity in AI research.

From Chengdu to Computer Science at Princeton

Fei-Fei Li was born in Beijing, China, in 1976. She grew up in Chengdu, the capital of China's Sichuan Province, which is known for its spicy food and natural beauty. As a child, Li would often take long walks with her father, Li Shun. He worked in the computer department of a chemical plant and instilled in Li a lifelong love for science and nature. She later recalled the feelings of excitement she had as a child exploring the mountains around Chengdu with her father. Spotting an unfamiliar butterfly or an exotic type of flower gave her the sort of thrill she later got from doing scientific research. Her mother's background as a school-teacher also helped Li become passionate about learning.

When Li was fifteen, her family moved to the United States. They settled in Parsippany, New Jersey, where her father repaired cameras and later ran a dry-cleaning business. Li worked hard to learn English and soon became an excellent student, earning a scholarship to Princeton University. There she studied physics, computer science, and engineering, while also helping out on weekends at her parents' dry-cleaning store.

Fei-Fei Li developed a method for training computers to see. Her dataset project, called ImageNet, ended up revolutionizing AI by teaching it to see a more complex world.

> "One of the inspirations in my early days of trying to unlock this visual intelligence problem was to look at how our visual semantic space is structured. There are so many tens and thousands, millions, of objects in the world. How are they organized [in the human mind]?"[5]
>
> —Fei-Fei Li

In the early 2000s, Li enrolled at the California Institute of Technology in Pasadena, California, to pursue her PhD in computer science and neuroscience. At the university, she became fixated on an idea that would change her life. She realized that progress in AI would hinge on training it with much larger datasets than had yet been employed. Li found support for her idea in a paper by the psychologist and neuroscientist Irving Biederman on how humans categorize and understand images. According to Biederman's estimates, the average person can recognize about thirty thousand different types of objects. "One of the inspirations in my early days of trying to unlock this visual intelligence problem was to look at how our visual semantic space is structured," says Li. "There are so many tens and thousands, millions, of objects in the world. How are they organized [in the human mind]? Are they organized by alphabet, or by size or colors?"[5] She began to get a glimmer of how she might approach the problem by adapting it to AI capabilities.

The ImageNet Project

Li soon got the opportunity to test her insight about training AI. In 2007 she became an assistant professor at Princeton and began to investigate a revolutionary database created there in the mid-1990s called WordNet. It arranged words not as separate definitions but as interconnected concepts in a semantic network. WordNet was based on a basic question about human language: How do people organize and access the huge vocabulary stored in their minds? Thinking about words is not simply remembering definitions. It also means going through a rich network of associations—for instance, *cars* are a kind of vehicle, *wheels* are parts of cars, and *automobile* and *car* mean pretty much the same thing. Li saw that this network idea could also be used to organize images. As she explains:

> In a dictionary, *apple* and *appliance* are very close together, but in real life an *apple* and a *pear* are much closer to each other. The apple and pear [are] both fruits, whereas an appliance belongs to a whole different family of objects. I made a connection in my head. This could be the way visual concepts are organized, because apples and pears are much more connected than apples and washing machines.[6]

Li began assembling a team to compile a massive database of images, like the word database in WordNet but much larger. For two years, Li and her team of academic colleagues and graduate students combed the internet for images of everything from aardvarks to zippers. A simple concept like a chair called for pictures of a recliner, rocker, beanbag chair, wingback chair, high chair, swivel chair, and dozens more. When the task became too unwieldy, Li turned to Amazon Mechanical Turk, a crowdsourcing platform of nearly fifty thousand workers worldwide, to filter and

An Inspirational Teacher

Those who achieve great things often attribute their success to a single inspirational mentor. For Fei-Fei Li, that mentor was Bob Sabella, a math teacher at her high school in New Jersey. Li faced many challenges during her first years in America, such as learning English and adapting to new surroundings and a different culture. Sabella was quick to recognize fifteen-year-old Li's potential. He worked past the language barrier to encourage her love for literature, science, and mathematics. He created a special one-person Advanced Placement math class for Li during lunch hour, since the school did not offer one. Li also credits Sabella for fostering her strong sense of curiosity about the world, which led to her innovative work in AI.

After Li graduated high school, Sabella continued to advise her on her academic career. They remained friends, and he attended her wedding. When Li interacts with young people today, she instantly recalls Sabella's foundational influence on her. She says, "My experience learning from Bob comes flooding back during moments like these—Mr. Sabella, as I knew him back then—and how deep an imprint a teacher can leave on the trajectory of youth. We're entrusted with sharing a special kind of joy—the thrill of knowledge, the rush of new possibilities."

Quoted in Jordan Jacobs, "Fei-Fei Li's Inspiring Memoir About Her Remarkable Life and the Rise of AI," *Radical Blog*, Radical Ventures, November 12, 2023. https://radical.vc.

label the images. Eventually, the team's database, which Li called ImageNet, grew to include twenty-two thousand classes of objects and more than 15 million images with annotations.

Experts in AI questioned the need for such a gigantic database, but Li thought it was necessary for machine learning. Although human children can learn rules based on a few examples, computers require large amounts of data to employ rules and categorize things. And while the term *big data* arose in the 1990s, it was Li's breakthrough research that helped make it a crucial consideration in the 2010s. "Pre-ImageNet, people did not believe in data," Li says. "Everyone was working on completely different paradigms [basic models] in AI with a tiny bit of data."[7]

Li hoped that others would build on her team's efforts, and she was not disappointed. In 2009 she joined the faculty at Stanford University in Palo Alto, California, becoming director of the Stanford Artificial Intelligence Lab (SAIL). She launched an annual contest in which researchers used ImageNet to train their computer

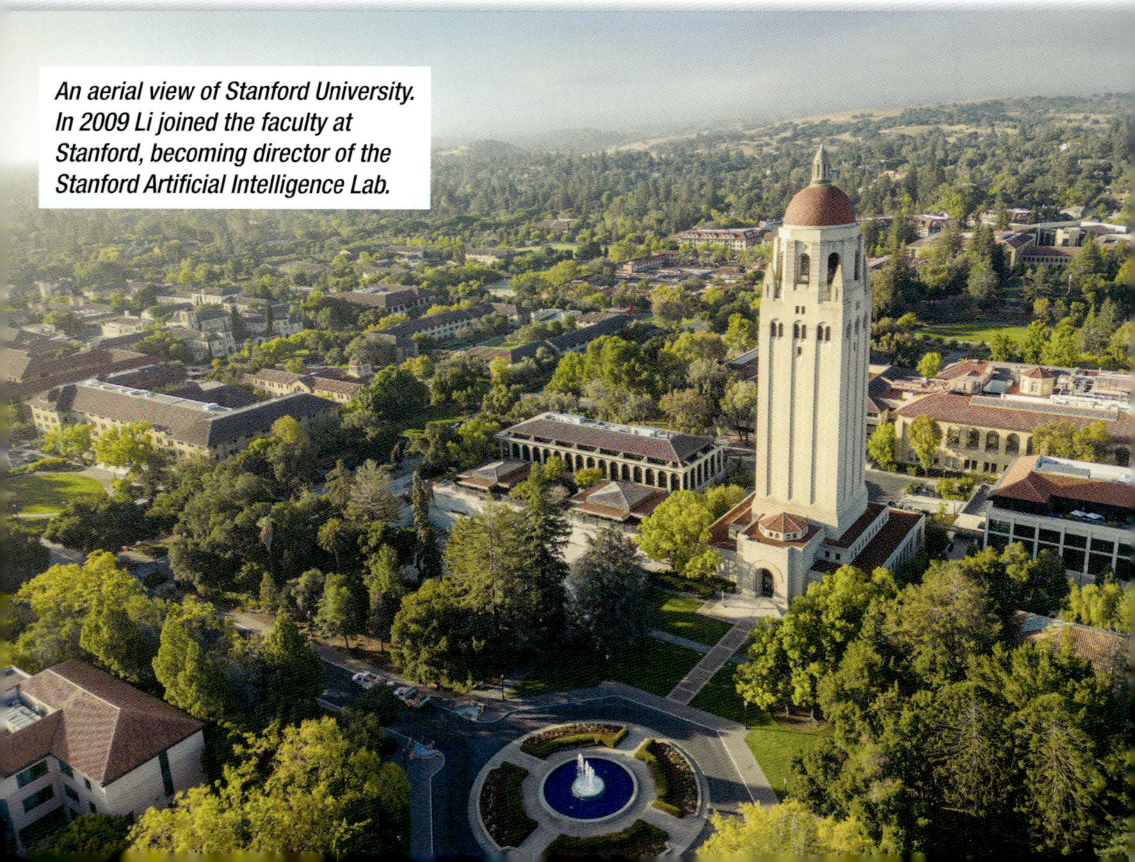

An aerial view of Stanford University. In 2009 Li joined the faculty at Stanford, becoming director of the Stanford Artificial Intelligence Lab.

systems in machine learning. The winner in 2012 was the University of Toronto lab run by AI visionary Geoffrey Hinton. The Toronto lab's AlexNet leaped far beyond past winners to set new standards in computer vision and identifying objects. Hinton's group combined brain-like neural networks with Li's database to achieve the lab's breakthrough. The result was a boom in deep learning and computer vision that continues today. It led not only to LLMs and AI platforms like ChatGPT and Claude but also to everyday applications such as self-driving cars and facial recognition software. As for Li, her pioneering work on ImageNet has earned her the title Godmother of AI.

Pushing for Diversity and Inclusion in AI Research

Li's rise to prominence in the field of AI gave her the opportunity to advocate for values in which she believes. Chief among these is the importance of diversity and inclusion in AI research and scientific work in general. She is especially focused on opportunities for those who traditionally have been underrepresented in science classes and research groups. In 2017 she started AI4ALL (AI for All), a nonprofit educational program that offers Stanford's SAIL courses to students in secondary school. AI4ALL seeks to encourage young women and minority students to pursue computer science as a career.

Another reason that Li launched AI4ALL was the alarming amount of bias that crept into the ImageNet database from online sources. Reports of Blacks and other minorities being mislabeled and certain jobs assigned only to White males made her rethink the project's parameters. "When we were pioneering ImageNet, we didn't know nearly as much as we know today," Li says. "We have massively evolved since. But if there are things we didn't do well, we have to fix them. . . . I'm cautiously optimistic that there are both technical solutions and governance solutions, as well as market demands to be better and better."[8]

Li is notably shy and reluctant to talk about herself. However, during the COVID-19 pandemic, she began working on a book about how she developed ImageNet. John Etchemendy, a philosopher and colleague of Li's at Stanford, read the manuscript and strongly suggested she start over. He wanted her to include more of her personal story: her childhood in China, her immigration to the United States, her struggles to learn English as a teenager, and the obstacles she had to overcome to achieve her lofty position in AI research. Her revised book, *The Worlds I See: Curiosity, Exploration, and Discovery at the Dawn of AI*, was widely praised for its focus on an immigrant woman's journey as a scientist. "[Etchemendy] said there's plenty of technical people who can read an AI book," says Li. "But I was missing an opportunity to tell all the young immigrants, women, and people of diverse backgrounds to understand that they can actually do AI, too."[9]

"There's plenty of technical people who can read an AI book. But I was missing an opportunity to tell all the young immigrants, women, and people of diverse backgrounds to understand that they can actually do AI, too."[9]

—Fei-Fei Li

Bringing Humanity to AI Systems

Another of Li's concerns is to ensure that AI will serve human needs and values. She founded and is codirector of the Institute for Human-Centered AI (HAI) at Stanford. Its mission is to advance AI through research, education, policy, and practice. HAI promotes a human-centered framework, guiding AI technology to improve the human condition. As Li notes, AI is capable of generating huge increases in productivity, but it may also eliminate jobs on a large scale. It can help communities interact and work together, but it can also introduce bias and misinformation. Li believes that societies need to consider ways to combat bias, protect the dignity of workers, and share the prosperity that AI is likely to bring.

Education will be vital to the cause of human-centered AI. HAI creates student affinity groups at Stanford so that members can learn about AI, share ideas, and debate its most beneficial uses. Li hopes that AI education will become standard for younger students as well. "We're hearing about some schoolteachers absolutely banning AI," she says. "But we also see some children starting to use AI in a responsible way and learning to take advantage of this tool. And the difference between those who understand how to use AI and those who do not is going to have extremely profound downstream effects."[10]

Some experts in the field believe AI is on a path to escaping human control. In fact, Geoffrey Hinton has said there is a significant chance that advanced AI will lead to human extinction. However, Li does not share his gloomy outlook. While she admits there are likely to be bad effects, she believes they can be managed. In her view, the doomsayers too often exaggerate the dangers of AI's evolution, making people unnecessarily fearful of the technology.

Working on AI's Next Frontier

Li's restless intelligence is constantly seeking new applications for AI. Her latest project promises a new frontier in which AI unfolds not on computer or phone screens but as a complex model. This

An Expert Opinion on a Bill About AI Safety

In the fierce debate over an AI safety bill in California, it is not surprising that the Godmother of AI was expected to weigh in. The bill, called the Safe and Secure Innovation for Frontier Artificial Intelligence Models Act (SB-1047), was intended to protect against catastrophic risks from future AI systems. These might include anything from AI-based cyberattacks on infrastructure to mass casualties from a rogue AI-controlled weapon of mass destruction. California State Senator Scott Wiener, who introduced the bill, sought to require more transparency of the largest AI companies, including reports on safety and security protocols whenever a safety issue arose.

As one of the world's best-known experts on AI, Fei-Fei Li had a keen interest in the debate. While she strongly supports AI safety measures, she believed the bill's requirements would do more harm than good. "If passed into law, SB-1047 will harm our budding AI ecosystem," she said. "[It] will unnecessarily penalize developers, stifle our open-source community, and hamstring academic AI research, all while failing to address the very real issues it was authored to solve." In October 2024 California Governor Gavin Newsom vetoed the bill, while signing a number of narrower, more targeted bills on AI safety.

Fei-Fei Li, "'The Godmother of AI' Says California's Well-Intended AI Bill Will Harm the U.S. Ecosystem," *Fortune*, August 6, 2024. https://fortune.com.

"Picture a utility operator facing wildfire-level winds. A spatially aware AI doesn't wait for a prompt. It forecasts how conditions will evolve, reroutes power, dispatches a drone to inspect a transformer predicted to fail, and alerts first responders—*before* anything burns."[11]

—Samantha Walravens, tech journalist

model is based on awareness (by way of machine learning) of conditions in the real world. In 2024 Li cofounded World Labs, a company that uses so-called spatial intelligence to create three-dimensional virtual worlds. Its AI systems use prompts from text, images, and videos in building incredibly detailed virtual environments. Moreover, these models can understand the elements of reality, such as objects, motion, barriers, and cause and effect. Spatial intelligence takes computer vision to a new level of humanlike awareness.

World Labs' first product is called Marble. It is designed to generate and simulate 3-D worlds and enable users to interact with them. The virtual worlds are created with fairly simple image or text prompts. Once created, a 3-D world can analyze complex scenarios and respond with appropriate actions. It can also be edited, expanded, or combined with other worlds. Tech journalist Samantha Walravens believes that Marble could have game-changing applications for real-world engineers. "Picture a utility operator facing wildfire-level winds," says Walravens. "A spatially aware AI doesn't wait for a prompt. It forecasts how conditions will evolve, reroutes power, dispatches a drone to inspect a transformer predicted to fail, and alerts first responders—*before* anything burns."[11]

Just as ImageNet employed vast amounts of data to help computers see, Marble and other World Labs products intend to make computers aware of real-world environments. In other words, Li and her colleagues are trying to instill computers with humanlike intuition on how the world actually works. This aligns with Li's mission to make AI more human friendly. "For the first time in history, we're poised to build machines that we can rely on as true partners in the greatest challenges we face," she says. "We're on the cusp of technology that elevates the aspects of life we care about most. This is a vision of deeper, richer, more empowered lives."[12]

CHAPTER TWO

Demis Hassabis: Using AI to Predict Biological Structures

The winner of the 2024 Nobel Prize in Chemistry learned about the award just minutes before it was announced to the world. Demis Hassabis, a British computer scientist and CEO of Google's DeepMind AI lab, had already missed several calls from the Nobel committee. They went to his wife instead, who did not recognize the Swedish number on Skype. However, when he finally received the committee's call, Hassabis was not exactly shocked. He already suspected that he and two of his DeepMind colleagues were likely to win. The day before, AI expert Geoffrey Hinton and an associate had shared the Nobel Prize in Physics for their work on machine learning. The 2024 Nobel Prizes were the first in history to recognize AI's contribution to science. The committee seemed to be announcing that AI solutions had become the dominant theme in scientific circles.

Hassabis's award recognized his achievement in biochemistry. He and his team had used DeepMind's AI resources to predict the structures of DNA proteins much more rapidly than was previously possible. Their breakthrough promised to boost research into new vaccines, cancer cures, human longevity, and many other areas. As one of the world's chief innovators in AI, Hassabis views the prestigious award as more of a beginning than an ultimate goal. Like other leaders in AI, he holds out great hopes for artificial general intelligence, in which AI systems will potentially overtake human capabilities. "I think some of the biggest problems that we face today as a society, whether that's climate or disease, will be helped by AI solutions," Hassabis says. "I'd be very worried about society today if I didn't know that something as transformative as AI was coming down the line."[13]

An Early Success in Computer Games

Demis Hassabis was born in London, England, in 1976. His father, Costas, is from Cyprus, an island in the Mediterranean, while his mother, Angela, comes from Singapore. By age four Hassabis was a chess prodigy. He went on to win tournaments playing against adults while only thirteen. He used his chess winnings to purchase his first computer. Learning programming from technical books, he was soon able to write rudimentary AI programs for board games. He excelled in school at Christ's College in London, finishing his A-level exams at sixteen, two years early.

> **"I think some of the biggest problems that we face today as a society, whether that's climate or disease, will be helped by AI solutions. I'd be very worried about society today if I didn't know that something as transformative as AI was coming down the line."[13]**
>
> —Demis Hassabis

Too young to begin his college career at Cambridge University, Hassabis spent a year off codesigning and programming an AI simulation game at Bullfrog Productions. The game, called *Theme Park*, involved players building and operating amusement parks. It soon became a huge hit, selling several million copies and starting a whole new genre of so-called sandbox games. Working on the game showed Hassabis the importance of combining creativity with technical skill. "Many children start by playing games, like I did," he says, "then get into programming, and then use this incredible tool, the computer, to create things."[14] His early success with computer games convinced him that his future lay in AI.

After a brilliant collegiate career at the University of Cambridge, in which he earned an advanced degree in computer science, Hassabis launched his own video game company, Elixir Studios. At Elixir he produced a number of award-winning games based on cutting edge AI concepts. He then returned to academia, earning his PhD

Demis Hassabis, seen at the Time 100 Gala in 2017, is one of the world's chief innovators in AI. A British computer scientist, Hassabis is the CEO of Google's DeepMind AI lab.

in cognitive neuroscience at University College London. His studies in neuroscience and brain function—especially the mechanisms involved in imagination, memory, and amnesia—played a large part in his later work in AI.

Founding DeepMind for Cutting Edge AI Research

In 2010 Hassabis joined with his colleague Shane Legg and childhood friend Mustafa Suleyman to create DeepMind, an ambitious London-based start-up focused on AI research. Its goal, as Hassabis has said, was to "solve intelligence and then use it to solve everything else."[15] The company's main research dealt

A Killer App in Wearable AI

As a world-class innovator in AI, Demis Hassabis has a résumé filled with major accomplishments, from AI systems that can defeat champion Go players to a Nobel Prize in Chemistry. One thing he has yet to develop, however, is a killer app or product that everyone simply has to have, like the iPhone or ChatGPT. But his new breakthrough in wearable AI may become the latest must-have product among tech enthusiasts and regular consumers alike.

The breakthrough is basically smart eyeglasses powered by Google's Gemini 3 AI. Hassabis admits that past attempts at smart specs have fallen short. But he believes that rapid improvements in AI will open a world of possibilities. "I think the killer app is a universal digital assistant that's with you, helping you in your everyday life, and available to you on any surface," says Hassabis, "on your computer, on your browser, on your phone, but also on devices like glasses when you're walking around the city. . . . I feel we finally got AI that is maybe powerful enough to make that a reality." Google DeepMind is planning to partner with eyewear firms Warby Parker and Gentle Monster as well as electronics powerhouse Samsung, with prototypes for the smart glasses due in 2026.

Quoted in Alex Kantrowitz, "Google DeepMind CEO Demis Hassabis on AI's Next Breakthroughs, What Counts as AGI, and Google's AI Glasses Bet," *Big Technology*, January 29, 2026. www.bigtechnology.com.

with deep reinforcement learning, in which AI systems not only attempt to make models of their digital world and discern patterns (as in deep learning) but also make active decisions and try to achieve goals.

DeepMind's founders flexed their research muscles by applying their insights to computer games. They trained an AI program to master a series of Atari video games by analyzing pixels on a computer screen to maximize its scores. Hassabis and his DeepMind colleagues were staking their company's claim as the leading new AI lab in the industry. In January 2014 Google showed its belief in DeepMind's potential by buying it for an estimated $500 million. The new partnership allowed DeepMind to use Google's data centers and massive computing power to train its AI algorithms.

In March 2016 DeepMind took aim at another deep learning milestone. Hassabis traveled to Seoul, South Korea, to watch the company's AI AlphaGo program take on the world champion in the ancient board game Go. Reputed to be the world's most complex game, Go requires a player to contemplate far more possible moves than chess. Lee Sedol, the Go champion, expected to win easily, but AlphaGo dominated the match four games to one. At one point, AlphaGo astonished online observers worldwide with an audacious move that showed unmistakable signs of creative intelligence. Standing backstage in Seoul, Hassabis was impressed. AlphaGo's easy victory convinced him that DeepMind was ready to tackle a major new challenge.

Using AI to Predict the Shapes of Proteins

Since his Cambridge days, Hassabis had been intrigued by biology, particularly proteins, which form the chemical basis for DNA and RNA, the building blocks of life. He recalled how a researcher friend, Tim Stevens, called proteins the molecular machines that make life work. Stevens was obsessed with the problem of how proteins are structured. Each protein generally consists of a strip of twenty amino acids, which is folded into a complex 3-D shape of swirls, twists, and curlicues. The structure of the folds determines a protein's function. The protein hemoglobin, for example, which transports oxygen around the body, is folded into the shape of a tiny pouch. Figuring out the folded shapes of proteins and what they do could lead to huge breakthroughs in molecular biology, such as new drugs, vaccines, and gene therapies.

Like moves in the game Go, each protein ribbon presents a vast number of possible folded structures—as many as 10^{300} possible forms. Current trial-and-error methods could take months or even years to identify the structure of one protein. Hassabis hoped that DeepMind's AI, with its ability to find and process alternatives at great speed, could solve the protein-folding problem. "I was thinking about what we had actually done with AlphaGo," he says. "We'd mimicked the intuition of incredible Go masters. I thought, if we can mimic the pinnacle of intuition in Go, then why couldn't we map that across to proteins?"[16] Soon after its AlphaGo success, DeepMind shifted its focus from games to biology. It hired several biologists and created a team of scientists from various disciplines to concentrate on protein folding.

Hassabis traveled to Seoul, South Korea, to watch his AI AlphaGo program take on the world champion Lee Sedol (pictured) in the ancient board game Go.

In 2020 Hassabis's team introduced AlphaFold2, a game-changing AI system to address the protein structure challenge. It was able to predict, with amazing speed and accuracy, the 3-D structure of proteins from their amino acid sequences. AlphaFold2 entered a global competition called Critical Assessment of Protein Structure Prediction (CASP) to test its capability against other computerized prediction methods. It swept the contest so thoroughly that biologists were astounded. AlphaFold2 predicted protein structures using deep-learning techniques that could tease out pairs of amino acids sitting side-by-side due to twisting folds in the sequence. Its accuracy had margins of error equal to the width of a single atom. As Stevens noted, "They wiped the floor with the CASP competition by bringing together all these things biologists have been pushing toward for decades and then just acing the AI."[17] At a time when approaches to protein folding seemed to have stalled, Hassabis and DeepMind had revitalized the discipline.

> **"[Hassabis's team] wiped the floor with the CASP competition by bringing together all these things biologists have been pushing toward for decades and then just acing the AI."[17]**
>
> —Tim Stevens, Cambridge University researcher and friend of Demis Hassabis

Rapid Adoption and a Nobel Prize

AlphaFold2 has met with widespread and rapid adoption since its release in 2021. Biologists were thrilled that it could reduce the time for predicting protein structures from months or years to mere minutes. According to the National Library of Medicine, more than ten thousand scientific papers cited AlphaFold's potential in the first two years it was available. By late 2025 the AlphaFold Protein Structure Database—which is available for free—had exploded from its original 360,000 entries to over 241 million. More than 3.3 million users from more than 190 countries have visited the database. The entire archive has been downloaded over twenty-three thousand times. Aside from its value in drug creation and disease analysis, AlphaFold has helped solve problems that have stumped biologists for decades. For example, two separate teams of researchers used the AI tool to investigate how sperm and egg cells merge. AlphaFold's analysis found that a complex of three proteins perched on the head of a sperm locks onto the surface of an egg cell during fertilization.

Working with AI Tools to Solve Disease

Fresh from his success with DeepMind's AlphaFold, Demis Hassabis wants to take AI-powered biological research to a whole new level. Isomorphic Labs, a spinoff from Google DeepMind, is focused on research into novel treatments for disease, including cancer and immune disorders. Hassabis's goal for the new venture is to perform most of the drug discovery process by computer, thus avoiding traditional labs that call for expensive biological materials, rigid safety rules, and long trial periods. Isomorphic has already signed partnerships for research with drugmaker giants Eli Lilly and Novartis. The deals could be worth billions if the AI drug-discovery methods work as promised. "This is the No. 1 most beneficial application of A.I. out there," Hassabis told *The New York Times*. "Our mission, one day, is to solve all disease."

Investors are anxious to get on board. In March 2025 Isomorphic raised $600 million from venture capital outfits, including Thrive Capital, which makes large wagers on AI-related businesses such as OpenAI. The investment money will enable Isomorphic to hire the best young researchers. "This will be one of the most consequential companies around," says Hassabis. "We want it to be a real powerhouse in the industry."

Quoted in Michael J. de la Merced, "Isomorphic Labs, Google's A.I. Drug Business, Raises $600 Million," *New York Times*, March 31, 2025. www.nytimes.com.

DeepMind's breakthroughs with the AlphaFold system brought Hassabis new recognition and many honors and awards. In March 2024 he was awarded a knighthood for his services to AI research. On October 9 that year, he learned at the last minute that he, his DeepMind colleague John Jumper, and University of Washington researcher David Baker had won the Nobel Prize in Chemistry. In a statement following the announcement, Hassabis said:

> Receiving the Nobel Prize is the honor of a lifetime. . . . I've dedicated my career to advancing AI because of its unparalleled potential to improve the lives of billions of people. AlphaFold has already been used by more than two million researchers to advance critical work, from enzyme design to drug discovery. I hope we'll look back on AlphaFold as the first proof point of AI's incredible potential to accelerate scientific discovery.[18]

In 2024 Hassabis won the Nobel Prize for Chemistry. He is seen here (far right) with Princess Madeleine at the Nobel banquet in Sweden after accepting his award.

Congratulatory messages poured in from around the world. Scientists who understood the protein folding problem could appreciate the magnitude of Hassabis's achievement. However, there were also dissenting views. Some experts thought the prize should have been for computer science, not chemistry. Others questioned whether future Nobel Prizes in scientific fields would simply become awards for AI research. Still, many saw his Nobel as validation of AI's power to transform scientific discovery. "The benefits of this research are remarkable," says Annette Doherty, president of Britain's Royal Society of Chemistry, "as we can all look forward to applications improving our health and wellbeing. I am sure that [the recipients'] work will prove as inspirational to future generations as the discoveries of their predecessors who have been awarded this most prestigious honour."[19]

Creating a Universal AI Agent with Project Astra

When Hassabis cofounded DeepMind, one of his long-term goals was to develop artificial general intelligence. With AlphaFold successfully launched, he was able to return to AGI-related work. In 2024 the DeepMind lab released its own LLM, called Gemini, which combined generative AI more closely with Google search. In a short time, Gemini has established itself as the cutting edge of search technology. In December 2025 Google signed a multiyear agreement to integrate Gemini AI and cloud technology into Apple iPhones, Macs, and other devices. For an annual fee of about $1 billion, Google DeepMind will use its advanced AI to make Siri, Apple's built-in virtual assistant, more task oriented and conversational. The deal also cemented DeepMind's position as the leading source of generative AI for smartphones, as it also provides the tech for Samsung's Galaxy AI.

Hassabis hopes that his next breakthrough on the road to AGI will come with Project Astra. It expands DeepMind's Gemini LLM into an AI assistant capable of seeing, hearing, and conversing about nearly anything related to human knowledge. Astra provides

> "We have theories about what kinds of capabilities these systems [like Astra] will have. . . . But at the end of the day, how it learns, what it picks up from the data, is part of the training of these systems. We don't program that in. It learns like a human being would learn."[20]
>
> —Demis Hassabis

search results and answers, but it also performs everyday tasks such as buying airline tickets and making schedule changes. This so-called universal AI agent promises to be a major milestone in AI and machine learning. In fact, Hassabis admits that he is often startled by Astra's spongelike hunger for data and knowledge. "We have theories about what kinds of capabilities these systems will have. That's obviously what we try to build into the architectures," he says. "But at the end of the day, how it learns, what it picks up from the data, is part of the training of these systems. We don't program that in. It learns like a human being would learn. So new capabilities or properties can emerge from that training situation."[20]

From childhood, Hassabis has loved computer games and the logic of AI systems. His early success in creating AI simulation games gave him experience in setting up virtual worlds. He used his Cambridge degree in computer science and interest in neuroscience to do groundbreaking research in AI. As cofounder of the British AI lab DeepMind, his work on deep reinforcement learning resulted in AlphaFold, a game-changing AI system able to predict the structure of folding proteins in record time. Today, with Gemini and Project Astra, Hassabis is striving to adapt advanced AI as a human-based tool in the modern world.

CHAPTER THREE

Sam Altman: A Vision for AI's Future

Sam Altman is a man in a hurry. As cofounder and CEO of OpenAI, which launched ChatGPT, he finds himself in the middle of a race to expand AI as an everyday tool. Despite lacking a degree in computer science—or any undergraduate degree—Altman has emerged as a leading figure in AI worldwide. His ability to articulate a vision for AI's future development makes him a go-to interview subject for journalists and tech writers. He mixes easily with heads of state, diplomats, economists, tech entrepreneurs, prize-winning scientists, and rookie researchers. His willingness to court controversy and speak his mind, including with rival CEOs and companies, keeps his name in the headlines.

Altman's aggressive leadership style tends to make enemies and cause day-to-day turbulence in the halls and labs of OpenAI. But no one denies his ability as a promoter, even a visionary, when it comes to the pursuit of AI as a powerful force for good. Jill Popelka, CEO of AI cybersecurity company Darktrace, does not envy Altman's unique position in the AI wars. "I can't really imagine what it's like to be in his shoes," she says, "to be the hero or the villain of this technology that everyone is talking about."[21]

"I can't really imagine what it's like to be in [Altman's] shoes, to be the hero or the villain of this technology that everyone is talking about."[21]

—Jill Popelka, CEO of AI cybersecurity company Darktrace

An Obsession with How a Machine Can Think

Samuel Altman was born in Chicago, Illinois, in 1985 and grew up in St. Louis, Missouri. His mother is a dermatologist, and his father, who died in 2018, was a real estate broker. From

Sam Altman is the cofounder and CEO of OpenAI, which created ChatGPT. Despite lacking an undergraduate degree, he has emerged as a leading figure in AI worldwide.

childhood, Altman always had a strong curiosity about how the world works. At age eight, when his parents gave him an Apple Macintosh computer, he did more than play with it. He took it apart, rebuilt it piece by piece, and set about learning to code. One question drove his obsession with the computer: How does it think? Exploring that question would shape his future in the worlds of business and science.

Altman was eleven when he entered John Burroughs School, a prestigious prep school in a suburb of St. Louis. The school encouraged creative thinking and ambitious plans. This led Altman to the West Coast, where he studied computer science at Stanford University in Palo Alto, California. In the 1990s and early 2000s, Stanford harbored some of the greatest minds in computer technology and other scientific fields, including both teachers and students. The university's Gates Computer Science Building became a rich source for producing innovators in nearby Silicon Valley. Among them were graduate students like Larry Page and Sergey Brin, who went on to develop a revolutionary search engine called

Google. Altman was stirred by the tech fervor at Stanford, but restless ambition made him abandon his academic career without getting his diploma. He had big ideas of his own that he was determined to pursue.

Becoming a Tech Entrepreneur

In 2005 nineteen-year-old Altman dropped out of Stanford to found Loopt, a social networking app company. As CEO, he was able to raise more than $30 million from tech investors to get the app up and running. Loopt allowed users to map their social circle by sharing their location with friends in real time. It also helped them locate nearby sites and businesses and integrate messages with Twitter, Facebook, and other platforms. Although it fell short of mass adoption, Loopt was acquired for $43 million by a financial tech firm in Austin, Texas, in 2012. It also provided Altman invaluable experience in the rough-and-tumble world of start-up companies.

Years later, during an appearance at John Burroughs School, Altman told the students he never regretted his decision to leave college early. "This was not like my life plan, but this other opportunity came up," he said. "It seemed like a really fun thing to try, and importantly I realized that I could go back. I think that's the key to most risk. Most things are not a one-way door. You can try something, [and] if it doesn't work out, you can undo it."[22]

> **"This was not like my life plan, but this other opportunity came up. It seemed like a really fun thing to try, and importantly I realized that I could go back. I think that's the key to most risk. . . . You can try something, [and] if it doesn't work out, you can undo it."[22]**
>
> —Sam Altman

Altman next worked at Y Combinator (YC), a now legendary incubator for start-ups such as Airbnb, Dropbox, and Reddit. Having become a partner at YC in 2011, he had learned how to attract investors to *accelerate* (the industry buzzword) the growth of promising start-ups. It was a skill that would serve him well in AI and tech, where growth is viewed as essential. In 2014 Altman became president of YC, succeeding

A Breakup Between Two Tech Masterminds

A bitter feud between two billionaire entrepreneurs is likely to affect the future development of AI. OpenAI CEO Sam Altman and Tesla and SpaceX founder Elon Musk began as fellow investors in OpenAI in 2015. But Musk grew uncomfortable as rivals such as DeepMind seemed to be outperforming their start-up. He also felt that OpenAI was losing sight of its goal to develop AI for the benefit of humanity. Musk balked when Altman decided to abandon OpenAI's nonprofit status in order to raise cash and compete with other AI firms. In 2018 Musk told the other founders that OpenAI had to merge with Tesla and proceed under his control, or else he would leave. Despite worries about losing their richest investor, Altman and the others rejected Musk's proposal.

Since then, relations between the two tech titans have only gotten worse. Suits and countersuits have flown back and forth, with Musk claiming he was misled about the nonprofit direction of OpenAI. In 2023 Musk also started his own AI company, xAI. As for Altman, he admits he misjudged Musk. "For a long time, I looked up to him as an incredible hero, a great jewel for humanity," he says. "I have different feelings now."

Quoted in Grace Kay et al., "The History of Elon Musk and Sam Altman's Relationship and Feuds, Which Date Back to the Early Days of OpenAI," Business Insider, December 13, 2025. www.businessinsider.com.

the cofounder Paul Graham. He used his new expertise to start a growth-stage investment fund, YC Continuity, to help start-up companies grow even more rapidly. He also wanted to support ventures in diverse fields, such as biotech, space travel, and energy. In a blog post, Altman noted the extraordinary success of YC companies, whose total valuation had surpassed $65 billion.

OpenAI and a Focus on AI Research

Altman could no longer resist the siren call of AI. He understood that artificial intelligence would encompass virtually all the tech innovations, including search, text and image generation, and personal assistance. In 2015 he joined with a group of tech industry heavyweights to cofound OpenAI. Besides Altman, the ownership group included Elon Musk, Greg Brockman, Jessica Livingston (also from YC), Peter Thiel, Amazon Web Services, and Microsoft. Instead of a standard profit-driven strategy, Altman preferred to focus on cutting edge research into AI. The founding group agreed to the company's nonprofit approach, with a policy of transparent, ethical development of AI for the benefit of humanity. OpenAI raised an initial investment of over $1 billion. Meanwhile, Altman kept his position running YC, a decision that would soon lead to conflict.

As its name suggests, OpenAI began as an open-source research lab for AI. Open-source means that the underlying code or weights of its AI systems were made available to anyone. The company also shared its early innovations in machine learning and reinforcement learning.

In November 2022 OpenAI released, to little fanfare, its first AI chatbot, ChatGPT. It was trained on massive datasets culled from the internet, giving it incredible new capabilities. ChatGPT employed generative AI to create original content, such as text, images, video, music, and computer code. Word began to spread that finally a machine displayed the cleverness and reasoning prowess of a person. When tech enthusiasts learned of ChatGPT's ability to simulate human communication and create content, they began downloading the mobile app in

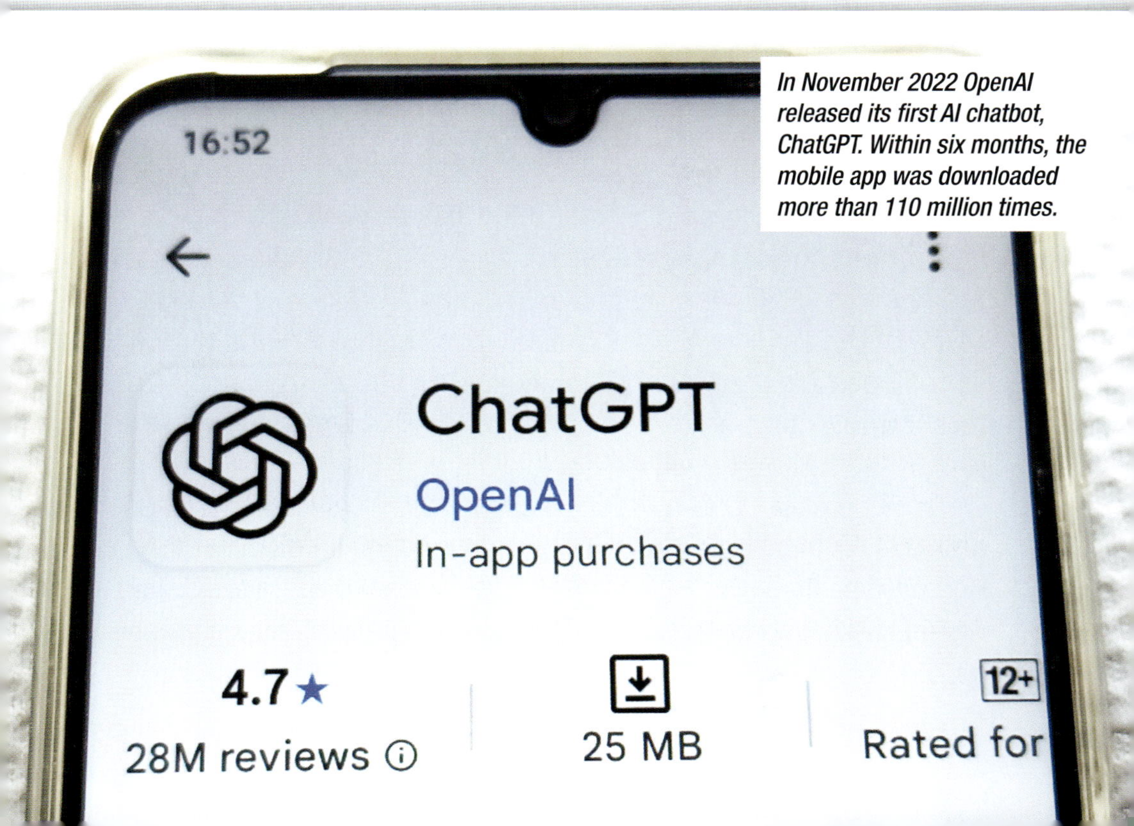

In November 2022 OpenAI released its first AI chatbot, ChatGPT. Within six months, the mobile app was downloaded more than 110 million times.

huge numbers. There were more than 480,000 downloads in its first five days and more than 110 million within six months. It is likely the fastest-growing tech product in history. And even better AI models would follow. Like the personal computer or smartphone, ChatGPT was a game-changing innovation, providing practical AI tools for the masses. As tech writer Philippe Hong observed in an analysis of Altman's career: "Imagine asking your phone to write an essay, plan a business, or crack a clever joke. Within seconds, it responds—intelligently and almost human-like. The technology amazes us, but here's the bigger question: *Who made this possible?* . . . These AI models weren't just tools; they were glimpses into the future of human–machine collaboration."[23]

A Clash over Leadership Style

In Altman's fervor to keep OpenAI at the forefront of AI research, he often displayed an abrasive style of leadership that could divide those around him. Controversy arose in 2019 when he announced the company would transition to a capped-profit format, replacing its former status as a nonprofit business. Investor returns were capped, or limited, to one hundred times their original investment—still a more-than-generous return. The move was aimed at attracting new investment money, since billionaire Elon Musk had quit the OpenAI board the year before and cut off his financial backing. Altman explained that the financial shift was necessary to deal with the staggering costs of developing AI, which were running into billions of dollars. Nonetheless, it irked employees and supporters who had relished the company's dedication to public benefit instead of profits.

In December 2023 a major clash erupted over Altman's leadership style. In a board decision that stunned the tech community, Altman was fired as OpenAI's CEO. The reasons for his removal were various and complicated. Some employees claimed he was verbally abusive, while others criticized his lack of transparency in making decisions. There were also questions about the company's breakneck pace of expansion and its erosion of safety measures in developing AI. Regardless of what led to his firing, tech experts feared for the future of OpenAI. Altman and Greg Brockman prepared to take their AI expertise to Microsoft, which was one of OpenAI's largest investors. Microsoft also promised to hire any of the company's employees willing to follow Altman out the door.

Then, in a shocking reversal just a few days later, Altman was reinstated as OpenAI's CEO. As it turned out, the vast majority of its employees—702 out of 770—had rallied to his defense. They threatened to leave unless Altman was restored to power. Their request for a new board of directors was also granted. Tech analysts saw the moves as proof of Altman's dominant position in the AI industry. Some also pointed out the danger in corporate power struggles having so much influence on AI's development. "Huge amounts of money—and huge egos—are in play," says Paul Barrett, deputy director of the Center for Business and Human Rights at New York University's business

AGI or Science Fiction?

Much ink and airtime have been spent on the delayed arrival of so-called artificial general intelligence, or AGI, a level at which AI can match or surpass human intelligence. But OpenAI CEO Sam Altman believes the term AGI is no longer very relevant as AI development surges ahead in many areas. And as a sort of honorary spokesperson for the AI industry, Altman's shifting opinion carries significant weight. "I think it's not a super useful term," he told financial reporters in August 2025. "I think the point of all this is it doesn't really matter and it's just this continuing [rapid growth] of model capability that we'll rely on for more and more things."

In reaching this conclusion, Altman is not alone. Many AI experts agree that the fuss about potential AGI distracts from more practical concerns. It can also raise unreasonable expectations about each new version of ChatGPT and its rival chatbots. "It drives funding and captures the public imagination," says Nick Patience, vice president of the AI-based Futurum Group, "but its vague, sci-fi definition often creates a fog of hype that obscures the real, tangible progress we're making in more specialized AI."

Quoted in Ryan Browne, "Sam Altman Now Says AGI, or Human-Level AI, Is 'Not a Super Useful Term'—and He's Not Alone," CNBC, August 11, 2025. www.cnbc.com.

school. "Judgments about when unpredictable AI systems are safe to be released to the public should not be governed by these factors."[24]

Controversy over Later Models of ChatGPT

Although ChatGPT remains the most popular chatbot by market share, the rollout of later models has been met with criticism and controversy. For example, GPT-5, released in August 2025, was hyped as a major upgrade to ChatGPT. It was promoted as having PhD-level intelligence and prodigious skills in coding. Its ability to route queries to different models automatically not only provided users a smoother experience but also saved money and energy. But soon after its release, complaints rained down from online user groups such as Reddit. The new model made errors, tended to hallucinate, and seemed sluggish in its responses. Many users missed the overall feel of the previous version, GPT-4o. "I've been trying GPT5 for a few days now," wrote one disgruntled member. "Even after customizing instructions, it still doesn't feel the same. It's more technical, more generalized, and honestly feels emotionally distant."[25] Some users reacted as if they had lost a close friend.

> **"Huge amounts of money—and huge egos—are in play. Judgments about when unpredictable AI systems are safe to be released to the public should not be governed by these factors."[24]**
>
> —Paul Barrett, deputy director of the Center for Business and Human Rights at New York University's business school

Altman admitted that GPT-5 included altered settings for safety reasons. The previous model had too readily agreed with nearly anything the user said, leading some users to depend on the chatbot for emotional support—sometimes to a dangerous degree. GPT-5 was programmed to take a cooler, more neutral approach to interactions. But as the backlash continued to grow, OpenAI made tweaks to the new model, such as patching a routing system that had led to errors and hallucinations. In addition, it gave the new model a warmer tone. "Rolling out to everyone is taking a bit longer," Altman wrote on X. "It's a massive change at big scale. We will continue to work to get things stable and will keep listening to

In 2025, ChatGPT-5 was released with altered settings to improve the previous model, which had too readily agreed with nearly anything the user said.

feedback. As we mentioned, we expected some bumpiness as we roll out so many things at once. But it was a little more bumpy than we hoped for!"[26]

OpenAI considered allowing users to switch back to the GPT-4o model if they chose. But in February 2026 it scrapped the 4o system altogether. The company reminded its critics that the year before it had published research showing how users were forming unhealthy emotional bonds with its chatbot models. Pattie Maes, a professor at the Massachusetts Institute of Technology who worked on the study, thinks the less flattering GPT-5 model was an improvement. "I personally think of that as a good thing, because GPT4o is also what led to delusions, bias reinforcement, etc.," says Maes. "But unfortunately many users like a model that tells them they are smart and amazing and that confirms their opinions and beliefs, even if [they are] wrong."[27]

A Code Red for AI Competition

Altman remains dedicated to keeping OpenAI ahead of its competitors in the race to achieve AGI. In December 2025 he alerted OpenAI employees to a code red, or emergency, situation due to intense competition from rival companies. He noted that tests of Google's Gemini 3 determined that it outperformed ChatGPT and other models on a variety of benchmarks. In response, Altman planned to spend over $1 trillion on infrastructure projects, such as building data centers for training new AI models. "This is the bet we are making," he told staffers, "and given our vantage point, we feel good about it."[28]

Altman has turned his childhood fascination with computers and gaming into a career based on innovations in technology. Leaving college at age nineteen, he became a successful tech entrepreneur. He leveraged his ideas about artificial intelligence to form OpenAI with a group of like-minded young industry stars. OpenAI's launch of ChatGPT and later models has propelled Altman and his company into the top rank of AI players worldwide. His pursuit of AGI is certain to be an exciting, and unpredictable, story in the years ahead.

CHAPTER FOUR

Daphne Koller: Bringing AI Solutions to Drug Development

As a successful immigrant in America, Israeli-born Daphne Koller wants other talented people to have the opportunities she has enjoyed. Her role in developing AI and starting two tech companies aligns with that of other immigrants who have been major AI innovators. Their importance is borne out by statistics. According to a 2023 analysis by the National Foundation for American Policy (NFAP), 77 percent of the top AI companies in the United States were founded or cofounded by immigrants or the children of immigrants. In fact, another NFAP study discovered that immigrants in the United States had launched more than half of the nation's billion-dollar companies, including several AI-based firms.

Koller has also pushed for improved education in science, technology, engineering, and mathematics (STEM) in her adopted country and increased focus on attracting women and minorities to scientific fields. But she worries about Trump administration policies that aim to trim science budgets and restrict immigration. Koller says:

> I would like nothing better than for the U.S. education system to really have the same emphasis on rigor and science and STEM so that we can train great scientists and engineers here. That would be incredible, but it doesn't happen magically. Even if that were ultimately the case, it's wonderful for a country to be in the unique position where it is the beacon, the magnet for the best and brightest from all over the world.[29]

Koller herself has proved how effective AI can be as a tool for education and research. Her start-up Coursera, with its AI-based platform, has provided targeted online learning assistance for millions of students. And her biomedicine company Insitro is employing machine learning to revolutionize drug discovery.

From Jerusalem to Computer Science at Stanford

Daphne Koller was born in Jerusalem, Israel, in 1968. Her parents both had academic careers, her father as a professor of botany and her mother as an English teacher. Their home was filled with books and lively conversation. However, while still a child and in spite of her parents' emphasis on learning, Koller found herself getting bored at school. She wanted to study poetry and ancient Greece, solve third-degree equations, and find out how mechanical things worked. Her classes seemed endless, moving far too slowly for her agile mind. Finally, at age thirteen, her parents allowed her to quit public school and enter Jerusalem's Hebrew University. Four years later, she had earned her bachelor's degree, and one year after that, her master's.

"Thanks to my family, I was able to bypass regular education and be myself. I have been very lucky. Since then I've been obsessed with one issue: How to make this possible for everyone?"[30]

—Daphne Koller

In 1993 Koller made another momentous decision. She left for the United States and was accepted into the computer science program at Stanford University, where she obtained a PhD. Following two years of postdoctoral research work at the University of California, Berkeley, she was hired as an assistant professor of computer science at Stanford. Her ability to find her own path influenced her thinking about improving education for others. "Thanks to my family, I was able to bypass regular education and be myself," she said later. "I have been very lucky. Since then I've been obsessed with one issue: How to make this possible for everyone?"[30]

At Stanford, Koller ran one of the world's top AI research labs. She did groundbreaking research in machine learning and probabilistic modeling. The latter seeks to quantify uncertainty and random patterns in data to predict the likelihood of various outcomes. It helps assess risk in fields such as engineering, finance, and machine learning. Koller also published research papers in the most prestigious journals. She was recognized as a leading computer scientist, winning a MacArthur Fellowship in 2004. She seemed to have the perfect academic career. But somehow, like that bored thirteen-year-old girl in a Jerusalem public school, she longed for a new challenge. She also wanted to justify so many people's belief in her. "When I did get that MacArthur Award, I actually felt very humbled and unworthy," she says. "And one might even say . . . that much of my career journey following the MacArthur Award was an attempt to kind of pay it back, to prove myself as having deserved that."[31]

An Online Education Venture with a Stanford Colleague

Koller found a like-minded colleague in the British-American computer scientist Andrew Ng. They shared a willingness to branch out and try new things. In particular, they both believed in seeking fresh ways to deliver education to the masses. The pair came up with an intriguing experiment: They offered several elite-level Stanford courses online, focusing on computer science and AI. There were no fees and no credits. Each course was interactive, with brief video lectures and forums for discussion. Students graded each other's quizzes and homework assignments, forming a sort of learning community. The courses were made available to anyone worldwide with an internet connection and a desire to learn. The response they got was immediate and overwhelming. "With no marketing, each of those courses had more than 100,000 learners," Koller recalls. "It was a wake-up call about the demand for this type of education around the world."[32] Koller and Ng pointed out that just one of their online courses drew more students than most professors teach in their entire lifetimes.

Andrew Ng (left) and Daphne Koller (right) cofounded Coursera. Over the years, Koller has proved how effective AI can be as a tool for education and research.

Koller and Ng realized that the technology existed to revolutionize education. They had also demonstrated that the demand for online learning, including high-level tech courses, was enormous. But when word got out about their plans to develop massive open online courses into a business, there was plenty of pushback from skeptical academics. Professors feared that the subject matter would be watered down and meaningful teacher-student interaction would disappear. They doubted that students would stay engaged. Some also worried about being replaced by AI.

Nonetheless, in January 2012 Koller and Ng took a leave of absence from Stanford to start Coursera, an online education company. It rapidly became a hit, serving about 4 million learners within its first two years. Coursera built partnerships with top universities, including Stanford, the University of Michigan, the University of Pennsylvania, Princeton, and Yale. Eventually, Coursera would enlist more than 350 universities and industry educators to provide courses and degree programs. With the cost of higher education and student debt loads skyrocketing, Coursera's free online classes and modest fees to obtain certificates of completion struck a chord with students in the United States and around the world. Many corporations and businesses set up agreements to use Coursera to train or reskill their personnel in tech and other areas. Courses could also be translated into different languages to make them more accessible. As online learning gained acceptance, a number of competitors arose, such as Udacity, 2U, and Knewton. In August 2016, when Koller stepped down as president of Coursera to pursue other interests, the company had served as many as 23 million registered learners with its menu of thirteen hundred courses.

Exploring Biomedicine with AI

Koller next became chief computing officer at Calico Life Sciences, a subsidiary of Alphabet, which owns Google. It was her first foray into biomedicine, a field that had long stirred her interest.

A Speedy Method to Update Technical Skills

Daphne Koller's innovative work at Coursera led to widespread acceptance for online education. Today it is an important tool to keep technical skills up-to-date. According to estimates by the World Economic Forum, more than 1 billion workers are in need of reskilling by 2030. Technical skills must be refreshed about every two years, especially in areas such as AI, machine learning, and data science.

As the pace of workplace change has increased, a new kind of AI-based education has emerged: micro-credentials. "Micro-credentials are short, skill-based credentials issued by universities, tech firms and online training organizations," says Venkatadri Marella of the Forbes Technology Council. "Unlike traditional degrees, they don't take years or heavy fees to pursue. Instead, they allow learners to attest proficiency in a single skill, optimizing big language models, developing AI-driven chatbots or applying machine vision to robotics, usually within weeks."

Most tech employers are open to this new method of targeted certification. It enables workers to gain new skills within weeks instead of pursuing a general two-year course or college major. A September 2024 Coursera survey of higher education leaders found widespread acceptance of micro-credentials. It is another example of how Koller's insights about educational needs have gained influence.

Venkatadri Marella, "How Micro-Credentials Are Shaping the Future of AI-Driven Learners," *Forbes*, September 24, 2025. www.forbes.com.

Research at Calico dealt with mechanisms of aging and age-related diseases. Its overall mission was finding ways to extend human life. Koller's innovation was to apply AI and machine learning principles to analyze the large-scale biological datasets used in the company's studies of aging.

Calico was notably reluctant to discuss its research publicly. Koller, much in demand for public appearances, often spoke at industry conferences while working at the company. She enjoyed describing how machine learning could unlock some of the mysteries of human aging. However, in doing so she also sometimes touched on the inner workings of the Calico lab. This might be why she parted ways with the company after only two years.

In 2018 Koller embarked on a more significant role in biomedicine and health care. She launched Insitro, a San Francisco–based company devoted to using machine learning tools and biological datasets to search for lifesaving drugs. The name comes from combining *in silico* (performing experiments by computer) and *in vitro* (performing experiments in the lab or test tube). Koller

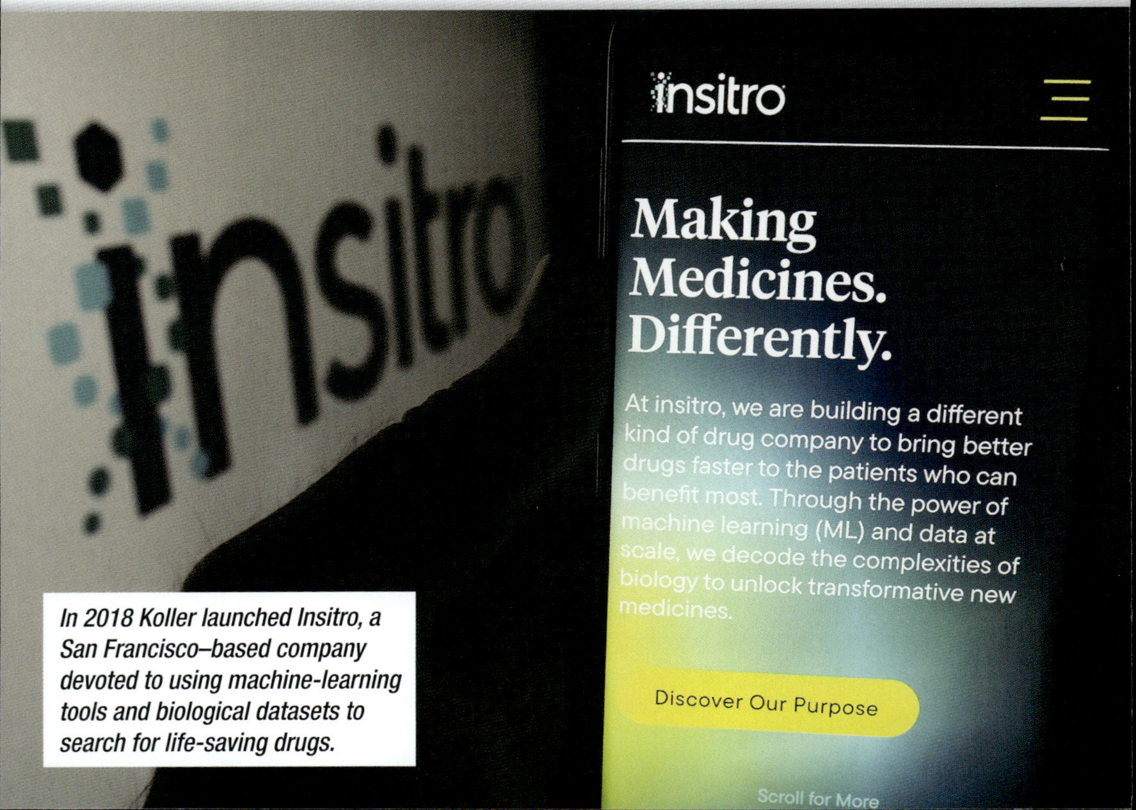

In 2018 Koller launched Insitro, a San Francisco–based company devoted to using machine-learning tools and biological datasets to search for life-saving drugs.

> "Koller's vision for Insitro reflects her belief that AI has the potential to transform not just industries, but lives."[33]
>
> —Justyna Zajac, tech journalist

was driven to start the company in order to apply advances in AI to help real people with rare diseases. Her prestige in the tech industry enabled her to raise $100 million in funding within the first six months. She also struck an early partnership with Gilead Sciences, a biopharmaceutical research company, to develop treatments for nonalcoholic steatohepatitis, a progressive liver disease that is a major cause of liver transplants. Insitro has also collaborated with Bristol Myers Squibb to search for next-generation therapies for amyotrophic lateral sclerosis, a degenerative nerve disease.

Insitro's whirlwind launch left Koller amazed at how quickly a personal vision could come together to help people in need. The ability to combine machine learning with large biological datasets had arrived at a critical time for patients. AI's predictive capabilities can identify which molecular pathways are most likely to produce effective drugs. It could reduce drug development time from years to months and drastically lower costs. "Koller's vision for Insitro reflects her belief that AI has the potential to transform not just industries, but lives," says tech journalist Justyna Zajac. "By using machine learning to analyze massive biological datasets, her team hopes to uncover patterns that human researchers might miss, opening the door to more effective and personalized medicines."[33]

Another challenge with the new company was getting computer scientists and medical researchers to communicate in helpful ways. At first, they were inclined to talk past each other. Each group had their own priorities, which often did not overlap. According to Koller:

> You can take the most sophisticated, best meaning scientists from either side and put them in the same room together and they might as well be speaking Thai and Swahili to each other. When you're an engineer, you're looking for the strongest, most consistent patterns that are going to

allow you to make predictions about a majority of cells or individuals. When you're a life scientist, oftentimes you're actually looking for the exceptions because those are the threads that can lead to new discoveries. So we've put in place a number of cultural elements and organizational elements to help people engage with each other openly, constructively and with respect.[34]

An Acquisition from Home

In January 2026 Insitro acquired CombinAbleAI, an AI-driven biotech company based in Israel. The deal added the Israeli company's expertise in antibody research to Insitro's TherML (therapeutic machine learning) platform. It also showed how cutting edge AI research has taken root all over the world. Where once Koller felt she had to leave Israel to pursue her tech dreams, today she can partner with a world-class Israeli lab for drug discovery.

Applying AI Skills to the Search for New Drugs

In a 2025 interview with *Fortune,* Daphne Koller discussed how she used AI and data collection to explore new techniques for drug discovery.

> My sort of journey into drug discovery came out of my actually early academic work at Stanford, where I'd been working at that intersection of biology and computer science and machine learning for quite a number of years prior to my detour at Coursera. . . . And so what I wanted to do was to bring the power of machine learning on the one hand, and equally importantly, the power of modern-day data collection from humans and different types of human-derived systems, bring that together to make as accurate a set of predictions as one can around whether a drug will have a benefit to a particular patient or not, and that requires having the right AI and having the right data to feed the AI, and those together are what is allowing us to create an engine that turns and makes more and more predictions about therapeutic hypotheses that we think are both novel and more likely to have a success in the clinic.

Quoted in Fortune Editors, "Why Two-Time Founder and MacArthur Genius Daphne Koller Sees a Future Where Humans 'Create Together with the Machines,'" *Fortune*, July 9, 2025. https://fortune.com.

“The goal is to use A.I. as a tool that empowers both students and teachers. The future of most human endeavors is a partnership between a human and a machine, and we need to teach our students toward that future.”[35]

—Daphne Koller

Koller has excelled at adapting AI and machine learning to benefit human society. From her research work as a professor at Stanford University to her role in founding two AI-driven companies, she has always pursued practical uses for AI. Foremost for her is the importance of education. In 2020 she and her husband, Dan Avida, started Engageli, her second AI-powered education company. It sought to boost student engagement in remote learning during the COVID-19 pandemic. It also aligned with her long-term aspirations for AI. “The goal is to use A.I. as a tool that empowers both students and teachers,” says Koller. “The future of most human endeavors is a partnership between a human and a machine, and we need to teach our students toward that future.”[35]

CHAPTER FIVE

Jensen Huang: Reinventing Computer Chips for the AI Age

On December 3, 2025, Nvidia CEO Jensen Huang had a bounce in his step as he exited his meetings with President Donald Trump and US lawmakers in Washington, DC. Dressed in a suit and tie rather than his usual black leather jacket and polo shirt, Huang felt like he had scored an important victory. For weeks he and other computer chipmakers had engaged in a fierce lobbying battle over the new national defense bill. Lawmakers had threatened to include a measure prohibiting sales of Nvidia's most advanced AI chips to China and other nations considered to be US adversaries. But Huang had reached a deal with the Trump administration to sell its second-best chips—the H200 model—to Chinese tech companies. In return, Nvidia agreed to give 25 percent of the sales revenue to the US government.

One month later, Huang had shifted his focus from negotiations to AI promotion. Onstage at the Consumer Electronics Show in Las Vegas, Nevada, he announced that AI was about to embark on a new era, one powered by chips that were not only faster but more specialized. He confirmed that Nvidia already was producing its next generation of AI processors, offering five times the computing power of previous models. Huang also touted the safety and transparency of Nvidia's AI data training for models like driverless cars. "Not only do we open-source the models, we also open-source the data that we use to train those models," he told the audience, "because only in that way can you truly trust how the models came to be."[36] In a rapidly changing AI landscape, Huang has become an authoritative guide to the future.

An Unsettled Childhood from Taiwan to the United States

Jensen (actually Jen-Hsun) Huang was born in Tainan, Taiwan, in 1963. His mother taught grade school, and his father was a chemical engineer. His childhood years were spent making frequent moves. At age five, his family left Taiwan for Thailand, but as the Vietnam War crept closer in neighboring Laos, the Huangs set their sights on a new life in America. In 1972 Jensen and his brother were the first to leave, flying to Tacoma, Washington, to live with their uncle. After a short stay, he shipped them off to what he thought was a religious school in Oneida, Kentucky, but it turned out to be a boarding school for troubled youths. As outsiders speaking broken English, the Huang brothers faced constant bullying from the other students. They endured ethnic slurs and threats with knives, but Jensen refused to be intimidated. He credits his time in Kentucky, with all its harrowing moments, for helping forge his strong character.

> "I washed the living daylights out of the dishes. And then they promoted me to busboy, I'm certain I was the best busboy Denny's ever had. No task was beneath me."[37]
>
> —Jensen Huang

The boys were in their teens when they finally reunited with their parents, who had immigrated to Portland, Oregon. There Jensen concentrated on his studies and found his first job as dishwasher at a local Denny's restaurant. "I washed the living daylights out of the dishes," he once told business students at Stanford University. "And then they promoted me to busboy, I'm certain I was the best busboy Denny's ever had. No task was beneath me."[37] To this day, he speaks with pride of having worked his way up to waiter. He insists that dealing with the stress of a crowded restaurant helped him learn how to keep his cool under pressure.

Huang went to Oregon State University, where he got his degree in electrical engineering and met Lori Mills, a fellow engineering student who would become his wife. In 1984 he moved to Silicon Valley in California to work at chipmaker Advanced Micro Devices. Although he spent only a year there, the company gave him an excellent introduction to microprocessors. He worked on designing these tiny electronic devices, which packed the function of a computer's central processing unit onto a single integrated circuit. From Advanced Micro Devices, he moved to LSI Logic, another cutting edge chipmaker. Its role in designing and manufacturing custom chips for Silicon Valley clients made it an important player in the microprocessor revolution. As an engineer at LSI Logic, Huang worked with some of the brightest people in

Nvidia CEO Jensen Huang arrives for meetings with President Donald Trump and US lawmakers to discuss a law prohibiting sales of the most advanced AI chips to US adversaries.

computer science. He also held positions in marketing and management. He was able to soak up technical knowledge every day that was much more valuable than his paycheck. During this time, Huang also earned his master's degree in electrical engineering from Stanford University.

A Partnership to Launch Nvidia

In late 1992 Huang was approached by two fellow Stanford graduates and chip designers who had worked at Sun Microsystems and IBM. Chris Malachowsky and Curtis Priem had learned about Huang's engineering skills firsthand when LSI Logic was supplying chip components to Sun Microsystems. The three met to discuss their business plans at a diner in San Jose—appropriately for Huang, it was a Denny's restaurant. They wanted to design and produce a better type of computer chip, capable of not only processing data but also creating vibrant new worlds in pixels. They believed they could revolutionize computer graphics and video games with 3-D graphics processing units (GPUs). Malachowsky would handle hardware design, Priem would deal with software architecture, and Huang would take charge of business decisions. On April 5, 1993, the trio founded Nvidia, basing its name on the Latin word *invidia*, meaning envy. The goal was to make the entire chip industry green with envy over their advanced products, an idea they stressed with their company's green logo.

In its early days, the new company flirted with failure. Other chipmakers remained skeptical about the need for a dedicated graphics chip. The trio's initial funding of $40,000 quickly evaporated, but they managed to secure a lifesaving $20 million investment from venture capital interests. Their first chip, the NV1, sank with little fanfare. Its modest follow-ups also disappointed. Half of the company's staff had to be let go, and there was talk of being thirty days away from going out of business. Still, Huang and his partners persevered. In 1999 they released their first true GPU, called the GeForce 256. It quickly proved to be a stunning milestone in computer graphics. "It was the first programmable

> "It was the first programmable GPU, capable of rendering 3D worlds with breathtaking detail and fluidity. Gamers gasped, developers marveled, and NVIDIA rocketed into the spotlight."[38]
>
> —Chandan Mishra, a marketing consultant in the tech industry

GPU, capable of rendering 3D worlds with breathtaking detail and fluidity," says Chandan Mishra, a marketing consultant in the tech industry. "Gamers gasped, developers marveled, and NVIDIA rocketed into the spotlight."[38]

A chief innovation of Nvidia's GPU was so-called parallel processing. It was able to break down larger tasks into smaller ones that ran simultaneously. By enabling devices to work on other functions faster, GeForce 256 provided graphics that were smoother and more realistic. Boosted by enthusiastic tech reviews and orders for the new chip, Nvidia's owners decided to make the company a publicly traded stock. When Nvidia's stock price soared past $100 a share, Huang celebrated by getting a tattoo of its logo on his shoulder. Years later, in 2023 the Denny's restaurant in San Jose put up a plaque honoring Huang and his partners. The inscription read: "The booth that launched a $1 trillion company."[39]

A Gift to the Future of AI

It might seem strange that in 2016 Jensen Huang would donate the world's first supercomputer to a small tech company, OpenAI, whose survival was uncertain. But Huang, like OpenAI investor Elon Musk, worried about tech giants such as Alphabet and Meta dominating AI research and hoarding young talent. He preferred tech companies to share AI innovations in an open process. So when Musk told Huang at a tech conference that he wanted Nvidia's supercomputer for OpenAI's research, Huang decided to oblige. Years later, he joked that since OpenAI was a nonprofit and could not pay for it, he almost nixed the deal. But Huang and Musk had friendly relations, and offering the supercomputer as a gift would make an important statement.

Huang famously boxed up the DGX-1 and drove it to OpenAI's cramped upstairs office in San Francisco. To commemorate the moment, he signed the supercomputer. "To Elon and the OpenAI Team!" he wrote in marker. "To the future of computing and humanity. I present you the world's first DGX1!" It was not just a gift to a struggling start-up but a gift to the world. OpenAI's engineers would use the supercomputer to experiment with massive datasets to train AI systems—and eventually produce ChatGPT.

Quoted in Paolo Confino, "Jensen Huang, Elon Musk, and OpenAI: The Story Behind the Hand Delivery of the First AI Supercomputer 'to Unlock the Powers of Superhuman Capabilities,'" *Fortune*, February 22, 2024. https://fortune.com.

In 1993 Huang, along with two fellow Standford graduates, founded Nvidia. The goal was to make the chip industry green with envy over their products, an idea they stressed with their company's green logo.

Embracing the Challenge of AI

The AI boom that erupted around 2010 dovetailed perfectly with Nvidia's GPU tech. Research scientists at Google and Stanford University were already using Nvidia GPUs to kick-start AI development. AI systems began to rely on neural networks, or computational models based on human brain processes. These networks required massive data inputs to learn how to recognize patterns and images. In 2012 Geoffrey Hinton and Alex Krizhevsky at the University of Toronto used the Nvidia GPUs to create their award-winning AlexNet image recognition network. For several years, Huang and his engineers at Nvidia had been adapting their GPUs for general data processing, and not just graphics production. Those GPUs were ideal for training neural networks and could do it one hundred times faster than traditional chips.

Around 2015, with AI technology about to explode, Huang went all in. He transformed Nvidia from a graphics-based business to a hardware company serving AI applications. Almost all the company's research and production going forward pivoted to focus on AI. Huang also played a key role in helping OpenAI develop its generative AI chatbot. In 2016 he hand delivered the first Nvidia supercomputer to the tiny OpenAI office in San Francisco. His gift is now considered a landmark moment, the beginning of the AI revolution. The Nvidia system he delivered included eight cutting edge GPUs, worth $300,000, which were eventually put to work training ChatGPT on huge datasets and LLMs. Upon ChatGPT's groundbreaking 2022 launch, tech experts marveled at the power and speed of Nvidia's GPUs in driving generative AI's many functions.

Huang realized that switching to practical AI-based computing would require a multiyear build-out, perhaps a decade or more. Speaking at a 2025 CEO tech summit in South Korea, he explained the economics behind the current boom. He called it a virtuous cycle, with large companies spending billions on AI-based infrastructure, such as chips and data centers, as the

Relieving the Data Center Strain on Power Grids

In the rollout of AI systems in the United States and worldwide, companies like Nvidia are spending billions to build large data centers to support AI. Data centers house many of the processes of digital infrastructure, such as data storage, cloud services, and AI functions. These data centers consume massive amounts of electricity, putting severe strains on local and regional power grids.

Nvidia CEO Jensen Huang acknowledges the AI-related problems of energy usage and sustainability, but he believes AI can also be the solution. "Data centers are already about 1–2% of global electricity consumption and that consumption is expected to continue to grow," says Huang. "This continued growth is not sustainable, neither for operating budgets nor for our planet. Accelerated computing is now the most sustainable way to advance computing." Nvidia achieves accelerated computing by combining its powerful GPUs with central processing units (CPUs) to manage complex computations more rapidly and efficiently. This can increase energy efficiency by twenty times or more over systems that rely on ordinary CPUs. According to Joshua Parker, senior director of corporate sustainability at Nvidia, "If you compare the energy efficiency for AI inference from eight years ago until today, it's 45,000 times more energy efficient."

Quoted in Jasmin Jessen, "NVIDIA: AI for Data Centre Energy Efficiency Improvements," *Energy Digital*, October 7, 2024. https://energydigital.com.

In 2025 Huang unveiled the GeForce RTX 5090, a GPU that contains 92 billion transistors and can perform 3,352 trillion AI operations per second.

technology improves. He described the cycle: "The AIs get better. More people use it. More people use it, it makes more profit, creates more factories, which allows us to create even better AIs, which allows more people to use it."[40]

A $5 Trillion Market Value

The ChatGPT launch also marked the beginning of Nvidia's meteoric ascent. Demand for the company's AI chips skyrocketed worldwide, giving it an 81 percent share of the chip market. One year after the launch, Nvidia reached a valuation of $1 trillion. By October 2025 its market value had risen to an eye-popping $5 trillion. Although its stock price backed off a bit, it remains the world's most valuable company. Huang's stake in the firm is reportedly worth about $160 billion, making him one of the ten richest people in the world.

With his urge to explore new areas of technology, Huang shows no signs of slowing down. Under his guidance, Nvidia continues to introduce a variety of state-of-the-art chips. At CES 2025, a consumer electronics show in Las Vegas, Nevada, he unveiled the GeForce RTX 5090, a GPU that contains 92 billion transistors and is able to perform 3,352 trillion AI operations per second. It optimizes performance for gaming and AI content creation. Huang referred to it simply as a beast. He also introduced Nvidia's Cosmos AI platform, which uses images and 3-D models of the real world to train humanoid robots, industrial bots, and self-driving cars. According to Huang, start-ups like Agility and Figure AI are already using Cosmos to teach their humanoid robots to make sophisticated movements and navigate in varied terrain. Cosmos models are trained on 20 million hours of video showing hand movements and the ways people walk. They can even teach warehouse robots how to prevent boxes from falling off a shelf. Nvidia also has produced Drive Thor, a so-called system-on-a-chip for autonomous vehicles that replaces multiple control systems with one master system. Drive Thor is capable of 2 quadrillion operations per second for AI-based decision-making in real time.

> "Most people think that [being a CEO] is about leading and being in command and being on top. None of that is true. . . . Most of the examples are making difficult decisions during very difficult times, it's mostly about sacrifice."[41]
>
> —Jensen Huang

As his company goes from one success to another, Jensen Huang often reflects on his journey from modest roots in Taiwan to CEO of a tech powerhouse. His ability to sense AI's amazing potential and take advantage of it with cutting edge computer chips has helped spread the technology worldwide. However, leading a firm with world-changing potential has also proved to be a constant source of stress and worry. He admits that he cannot watch a movie without thinking about some business decision or strategic choice. "Most people think that [being a CEO] is about leading and being in command and being on top," he says. "None of that is true. . . . Most of the examples are making difficult decisions during very difficult times, it's mostly about sacrifice."[41]

SOURCE NOTES

Introduction: Seeking AGI with DeepSeek

1. Quoted in Eduardo Baptista, "Who Is Liang Wenfeng, the Founder of DeepSeek?," Reuters, January 28, 2025. www.reuters.com.
2. J. Christopher Westland, "The Tragedy of Frank Rosenblatt and the Long Road to Neural Networks," LinkedIn, January 18, 2025. www.linkedin.com.
3. Sam Altman, "The Gentle Singularity," *Posthaven* (blog), June 10, 2025. https://blog.samaltman.com.

Chapter One: Fei-Fei Li: Teaching AI to See

4. Quoted in Timothy B. Lee, "How a Stubborn Computer Scientist Accidentally Launched the Deep Learning Boom," Ars Technica, November 11, 2024. https://arstechnica.com.
5. Quoted in Mishal Husain, "The Godmother of AI Didn't Expect It to Be This Massive," Bloomberg, November 21, 2025. www.bloomberg.com.
6. Quoted in Husain, "The Godmother of AI Didn't Expect It to Be This Massive."
7. Quoted in Lee, "How a Stubborn Computer Scientist Accidentally Launched the Deep Learning Boom."
8. Quoted in Steven Levy, "Fei-Fei Li Started an AI Revolution by Seeing Like an Algorithm," *Wired*, November 10, 2023. www.wired.com.
9. Quoted in Levy, "Fei-Fei Li Started an AI Revolution by Seeing Like an Algorithm."
10. Fei-Fei Li and Sara Frueh, "AI Is a Tool, and Its Values Are Human Values," *Issues in Science and Technology*, Spring 2024. https://issues.org.
11. Samantha Walravens, "AI Pioneer Fei-Fei Li Ushers In Next Frontier of Artificial Intelligence," *Forbes*, November 20, 2025. www.forbes.com.
12. Fei-Fei Li, "Spatial Intelligence Is AI's Next Frontier," *Time*, December 11, 2025. https://time.com.

Chapter Two: Demis Hassabis: Using AI to Predict Biological Structures

13. Quoted in Billy Perrigo, "Demis Hassabis Is Preparing for AI's Endgame," *Time*, April 16, 2025. https://time.com.

14. Quoted in Toronto Film School, "Game Changer: Sir Demis Hassabis and the Power of Video Game Design," October 18, 2024. www.torontofilmschool.ca.
15. Quoted in Steve Rose, "Demis Hassabis on Our AI Future: 'It'll Be 10 Times Bigger Than the Industrial Revolution—and Maybe 10 Times Faster," *The Guardian,* August 4, 2025. www.theguardian.com.
16. Quoted in Will Douglas Heaven, "This Is the Reason Demis Hassabis Started DeepMind," *MIT Technology Review*, February 23, 2022. www.technologyreview.com.
17. Quoted in Heaven, "This Is the Reason Demis Hassabis Started DeepMind."
18. Quoted in Albert Silver, "Demis Hassabis from DeepMind Wins Nobel in Chemistry," ChessBase, October 9, 2024. https://en.chessbase.com.
19. Quoted in Hannah Devlin, "Google DeepMind Scientists and Biochemist Win Nobel Chemistry Prize," *The Guardian* (US edition), October 9, 2024. www.theguardian.com.
20. Quoted in Scott Pelley et al., "What's Next in Artificial Intelligence According to a Tech Visionary Who May Hold the Cards to Our Future," CBS News, April 20, 2025. www.cbsnews.com.

Chapter Three:
Sam Altman: A Vision for AI's Future

21. Quoted in Sarah E. Needleman, "Tech CEOs Marvel—and Worry—About Sam Altman's Dizzying Race to Dominate AI," Business Insider, October 11, 2025. www.businessinsider.com.
22. Quoted in Paolo Confino, "Sam Altman Tells High Schoolers Dropping Out of College Wasn't a Big Deal," *Fortune*, September 19, 2024. https://fortune.com.
23. Philippe Hong, "Sam Altman: The Visionary Behind the AI Revolution," Founder Foundry, January 1, 2025. www.founderfoundry.com.
24. Quoted in Blake Montgomery, "'Huge Egos Are in Play': Behind the Firing and Rehiring of OpenAI's Sam Altman," *The Guardian* (US edition), November 23, 2023. www.theguardian.com.
25. Quoted in Will Knight, "OpenAI Scrambles to Update GPT-5 After Users Revolt," *Wired*, August 11, 2025. www.wired.com.
26. Quoted in Jason Nelson, "Bumps in the Machine: OpenAI's GPT-5 Rollout Stumbles into the Spotlight," Yahoo! Finance, August 8, 2025. https://finance.yahoo.com.
27. Quoted in Knight, "OpenAI Scrambles to Update GPT-5 After Users Revolt."
28. Quoted in Ashley Capoot, "OpenAI Is Under Pressure as Google, Anthropic Gain Ground," CNBC, December 2, 2025. www.cnbc.com.

Chapter Four:
Daphne Koller: Bringing AI Solutions to Drug Development

29. Quoted in Kate Zernike, "U.S. Scientists Warn That Trump's Cuts Will Set Off a Brain Drain," *New York Times*, June 3, 2025. www.nytimes.com.
30. Quoted in Facebook, "Women in History's Post," January 14, 2026. www.facebook.com.
31. Quoted in Diane Brady and Kristin Stoller, "MacArthur Genius Daphne Koller Sees a Future Where Humans 'Create Together with the Machines,'" *Fortune*, July 9, 2025. https://fortune.com.
32. Quoted in Goodnet, "How Coursera Is Revolutionizing the Way People Earn Degrees," November 1, 2018. www.goodnet.org.
33. Quoted in Steve Vassallo, "Using Artificial Intelligence to Hunt for New Drugs: Daphne Koller's Next Big Mission," *Forbes*, October 24, 2022. www.forbes.com.
34. Quoted in Matthew Perrone, "Better Drugs Through AI? Insitro CEO on What Machine Learning Can Teach Big Pharma," *Seattle (WA) Times*, December 2, 2024. www.seattletimes.com.
35. Quoted in Observer, "Daphne Koller Wants to Fix Drug Discovery with A.I.—and Real Data," September 18, 2025. https://observer.com.

Chapter Five:
Jensen Huang: Reinventing Computer Chips for the AI Age

36. Quoted in Stephen Nellis, "Nvidia CEO Huang Says Next Generation of Chips Is in Full Production," Reuters, January 6, 2026. www.reuters.com.
37. Quoted in Carmine Gallo, "The 9-Stage Hero's Journey for Leadership Storytelling," *Forbes*, January 15, 2026. www.forbes.com.
38. Chandan Mishra, "From Garage Tinkerers to Gaming Gods to AI: The Story of NVIDIA," Medium, April 15, 2024. https://mishrachandan.medium.com.
39. Quoted in Brian Caulfield, "NVIDIA Founder and CEO Jensen Huang Returns to Denny's Where NVIDIA Launched a Trillion-Dollar Vision," NVIDIA, September 26, 2023. https://blogs.nvidia.com.
40. Quoted in Tasmin Lockwood, "Nvidia CEO Jensen Huang Says AI Is in a 'Virtuous Cycle.' Here's What He Means," CNBC, October 31, 2025. www.cnbc.com.
41. Quoted in Eleanor Pringle, "Nvidia's Jensen Huang Says 'To Be a CEO Is a Lifetime of Sacrifice,' But His Parents Prepared Him for the 'Pain and Suffering' of Leadership," Yahoo! Finance, November 17, 2025. https://finance.yahoo.com.

FOR FURTHER RESEARCH

Books

Karen Hao, *Empire of AI: Dreams and Nightmares in Sam Altman's OpenAI*. Penguin, 2025.

Fei-Fei Li, *The Worlds I See: Curiosity, Exploration, and Discovery at the Dawn of AI*. Flatiron, 2023.

Sebastian Mallaby, *The Infinity Machine: Demis Hassabis, DeepMind, and the Quest for Superintelligence*. Penguin, 2026.

Parmy Olson, *Supremacy: AI, ChatGPT, and the Race That Will Change the World*. St. Martin's, 2024.

Stephen Witt, *The Thinking Machine: Jensen Huang, Nvidia, and the World's Most Coveted Microchip*. Viking, 2025.

Internet Sources

Diane Brady and Kristin Stoller, "MacArthur Genius Daphne Koller Sees a Future Where Humans 'Create Together with the Machines,'" *Fortune*, July 9, 2025. https://fortune.com.

Tom Huddleston Jr., "Nvidia Nearly Went Under 3 Years into Its Existence—Now It's the World's First-Ever $5 Trillion Company," CNBC, October 30, 2025. www.cnbc.com.

Mishal Husain, "The Godmother of AI Didn't Expect It to Be This Massive," Bloomberg, November 21, 2025. www.bloomberg.com.

Scott Pelley et al., "What's Next in Artificial Intelligence According to a Tech Visionary Who May Hold the Cards to Our Future," CBS News, April 20, 2025. www.cbsnews.com.

Zoë Schiffer and Will Knight, "Sam Altman Says ChatGPT Is on Track to Out-Talk Humanity," *Wired*, August 15, 2025. www.wired.com.

Organizations and Websites

Association for the Advancement of Artificial Intelligence (AAAI)
http://aaai.org
The AAAI is the premier scientific society dedicated to advancing the scientific understanding of the mechanisms underlying AI research and development. The group promotes research into responsible AI.

It also seeks to increase public understanding of AI. *AI Magazine*, the group's quarterly journal, helps members keep current with significant new research related to AI.

Brookings Institution
www.brookings.edu
The Brookings Institution is a nonprofit organization devoted to in-depth research to improve policy and governance at the local, national, and global levels. Its website features several articles about US policy toward the development of AI.

Center for AI Safety (CAIS)
https://safe.ai
The CAIS is a San Francisco–based nonprofit for AI-related research. Its mission is to reduce societal-scale risks associated with AI by conducting safety research, building the field of AI safety researchers, and advocating for safety standards. The CAIS website features articles about the hazards of irresponsible development of AI.

Hoover Institution
www.hoover.org
The Hoover Institution is an American public policy think tank that promotes economic liberty and prosperity, free enterprise, and limited government. Its website includes articles that examine AI's rapid rise, its enormous potential, and its possible drawbacks.

Stanford Institute for Human-Centered AI (HAI)
https://hai.stanford.edu
The HAI is an interdisciplinary institute established in 2019 to advance AI research, education, policy, and practice. It brings together thought leaders from academia, industry, and government to shape the development and responsible deployment of AI. The HAI website includes articles such as "AI Reveals How Brain Activity Unfolds over Time."

Union of Concerned Scientists (UCS)
https://support.ucs.org
The UCS is a member-supported nonprofit fighting for a safer, healthier, and more just world. The UCS was founded more than fifty years ago at the Massachusetts Institute of Technology. Its experts undertake rigorous, independent research, analyze pressing problems, and develop innovative solutions that are both equitable and cost effective.

INDEX

PICTURE CREDITS

Cover: Kyodonews/ZUMAPRESS/Newscom

5: VCG/Newscom
9: PA Images/Alamy Images
12: CravenA/Shutterstock
19: Debby Wong/Shutterstock
22: Imago/Alamy Images
24: TT News Agency/Alamy Images
28: Antonello Marangi/Shutterstock
31: Saku_rata160520/Shutterstock
35: Tada Images/Shutterstock
40: Associated Press
42: imageBROKER/Timon Schneider/Newscom
48: Associated Press
51: El editorial/Shutterstock
53: glen photo/Shutterstock